RECORD OF TRAINING LOG

CRANE

Date Training Log Started		Date Training Log Ended	
Name of Trainee:			
Address:			
Postcode:			
Telephone #:		**Mobile #:**	
Trainee's Signature			

The **Record of Training Logbook** assists in meeting legal obligations to provide a record of on-the-job training and supervision, in addition to any evidence of prior training received.

SUPERVISION: Training and Supervision must be done by a competent person. That is, a person deemed competent by virtue of having any of the following: Requisite license, qualifications, Certificate and years of relevant job experience. **Ensure that your supervisor signs the record of your work.**

© Safety Guides

RECORD OF TRAINING LOG

EMPLOYER INFORMATION

Name of Employer	Address	Telephone No.

RECORD OF TRAINING LOG

EMPLOYER INFORMATION

Name of Employer	Address	Telephone No.

RECORD OF TRAINING LOG

SUPERVISORS INFORMATION

Details	Supervisor 1	Supervisor 2	Supervisor 3
Name			
Address			
Signature			
Telephone#			
Certificate#			
State of Issue			
Issue Date			
Expiry Date			

RECORD OF TRAINING LOG

SUPERVISORS INFORMATION

Details	Supervisor 4	Supervisor 5	Supervisor 6
Name			
Address			
Signature			
Telephone#			
Certificate#			
State of Issue			
Issue Date			
Expiry Date			

RECORD OF TRAINING LOG

SUPERVISORS INFORMATION

Details	Supervisor 7	Supervisor 8	Supervisor 9
Name			
Address			
Signature			
Telephone#			
Certificate#			
State of Issue			
Issue Date			
Expiry Date			

RECORD OF TRAINING LOG

SUPERVISORS INFORMATION

Details	Supervisor 10	Supervisor 11	Supervisor 12
Name			
Address			
Signature			
Telephone#			
Certificate#			
State of Issue			
Issue Date			
Expiry Date			

COMPETENCY STANDARD
CRANE

COMPETENCY UNIT 1	COMPETENCY UNIT 2	COMPETENCY UNIT 3	COMPETENCY UNIT 4
Assess and secure equipment and work area	**Secure and transfer load**	**Set up and dismantle mobile R tower crane**	**Carry out special operations with mobile or tower cranes**
Conduct routine checks	Secure load	Plan assemble/disassemble	Travel crane
Plan work	Conduct trial lift	Set up crane	Carry out multiple crane lift
Check controls and lifting gear	Transfer load	Dismantle crane	
Shut down crane		Erect and dismantle tower crane	

*A competency unit is a wide module of work which has two segments: "Elements of competency" and their accompanying "performance criteria". Note all the competency units above to assist you in completing your logbook.

National Occupational Health and Safety Commission

RANGE STATEMENT FOR CRANES

All elements are to be satisfied in the normal workplace environment(s) or equivalent

EQUIPMENT RANGE INCLUDES

- tower cranes;

- derrick cranes;

- portal boom cranes;

- bridge or gantry cranes (cabin and remote over three operations);

- vehicle loading cranes (capacity 10 metre tonnes and above);

- non-slewing mobile cranes (greater than three tonnes capacity);

- slewing mobile cranes (up to 20 tonnes);

- slewing mobile cranes (up to 60 tonnes);

- slewing mobile cranes (up to 100 tonnes); and

- slewing mobile cranes (open/over 100 tonnes).

Date	Time Started	Time Ended	No. of hours	COMPETENCY UNIT Type of Work Performed	Machine/Equipment Details	Name of Trainer/ Supervisor
					Make:	
					Model:	Signature of Supervisor
					Serial No.:	
					Make:	
					Model:	Signature of Supervisor
					Serial No.:	
					Make:	
					Model:	Signature of Supervisor
					Serial No.:	

Date	Time Started	Time Ended	No. of hours	COMPETENCY UNIT Type of Work Performed	Machine/Equipment Details	Name of Trainer/ Supervisor
					Make:	
					Model:	Signature of Supervisor
					Serial No.:	
					Make:	
					Model:	Signature of Supervisor
					Serial No.:	
					Make:	
					Model:	Signature of Supervisor
					Serial No.:	

Date	Time Started	Time Ended	No. of hours	COMPETENCY UNIT Type of Work Performed	Machine/Equipment Details	Name of Trainer/Supervisor
					Make:	
					Model:	Signature of Supervisor
					Serial No.:	
					Make:	
					Model:	Signature of Supervisor
					Serial No.:	
					Make:	
					Model:	Signature of Supervisor
					Serial No.:	

Date	Time Started	Time Ended	No. of hours	COMPETENCY UNIT Type of Work Performed	Machine/Equipment Details	Name of Trainer/ Supervisor
					Make:	
					Model:	Signature of Supervisor
					Serial No.:	
					Make:	
					Model:	Signature of Supervisor
					Serial No.:	
					Make:	
					Model:	Signature of Supervisor
					Serial No.:	

Date	Time Started	Time Ended	No. of hours	COMPETENCY UNIT Type of Work Performed	Machine/Equipment Details	Name of Trainer/ Supervisor
					Make:	
					Model:	Signature of Supervisor
					Serial No.:	
					Make:	
					Model:	Signature of Supervisor
					Serial No.:	
					Make:	
					Model:	Signature of Supervisor
					Serial No.:	

Date	Time Started	Time Ended	No. of hours	COMPETENCY UNIT Type of Work Performed	Machine/Equipment Details	Name of Trainer/ Supervisor
					Make:	
					Model:	Signature of Supervisor
					Serial No.:	
					Make:	
					Model:	Signature of Supervisor
					Serial No.:	
					Make:	
					Model:	Signature of Supervisor
					Serial No.:	

Date	Time Started	Time Ended	No. of hours	COMPETENCY UNIT Type of Work Performed	Machine/Equipment Details	Name of Trainer/ Supervisor
					Make:	
					Model:	Signature of Supervisor
					Serial No.:	
					Make:	
					Model:	Signature of Supervisor
					Serial No.:	
					Make:	
					Model:	Signature of Supervisor
					Serial No.:	

Date	Time Started	Time Ended	No. of hours	COMPETENCY UNIT Type of Work Performed	Machine/Equipment Details		Name of Trainer/ Supervisor
					Make:		
					Model:		Signature of Supervisor
					Serial No.:		
					Make:		
					Model:		Signature of Supervisor
					Serial No.:		
					Make:		
					Model:		Signature of Supervisor
					Serial No.:		

Date	Time Started	Time Ended	No. of hours	COMPETENCY UNIT Type of Work Performed	Machine/Equipment Details	Name of Trainer/ Supervisor
					Make:	
					Model:	Signature of Supervisor
					Serial No.:	
					Make:	
					Model:	Signature of Supervisor
					Serial No.:	
					Make:	
					Model:	Signature of Supervisor
					Serial No.:	

Date	Time Started	Time Ended	No. of hours	COMPETENCY UNIT Type of Work Performed	Machine/Equipment Details	Name of Trainer/ Supervisor
					Make:	
					Model:	Signature of Supervisor
					Serial No.:	
					Make:	
					Model:	Signature of Supervisor
					Serial No.:	
					Make:	
					Model:	Signature of Supervisor
					Serial No.:	

Date	Time Started	Time Ended	No. of hours	COMPETENCY UNIT Type of Work Performed	Machine/Equipment Details	Name of Trainer/ Supervisor
					Make:	
					Model:	Signature of Supervisor
					Serial No.:	
					Make:	
					Model:	Signature of Supervisor
					Serial No.:	
					Make:	
					Model:	Signature of Supervisor
					Serial No.:	

Date	Time Started	Time Ended	No. of hours	COMPETENCY UNIT Type of Work Performed	Machine/Equipment Details		Name of Trainer/ Supervisor
					Make:		
					Model:		**Signature of Supervisor**
					Serial No.:		
					Make:		
					Model:		**Signature of Supervisor**
					Serial No.:		
					Make:		
					Model:		**Signature of Supervisor**
					Serial No.:		

Date	Time Started	Time Ended	No. of hours	COMPETENCY UNIT Type of Work Performed	Machine/Equipment Details	Name of Trainer/ Supervisor
					Make:	
					Model:	Signature of Supervisor
					Serial No.:	
					Make:	
					Model:	Signature of Supervisor
					Serial No.:	
					Make:	
					Model:	Signature of Supervisor
					Serial No.:	

Date	Time Started	Time Ended	No. of hours	COMPETENCY UNIT Type of Work Performed	Machine/Equipment Details	Name of Trainer/ Supervisor
					Make:	
					Model:	**Signature of Supervisor**
					Serial No.:	
					Make:	
					Model:	**Signature of Supervisor**
					Serial No.:	
					Make:	
					Model:	**Signature of Supervisor**
					Serial No.:	

Date	Time Started	Time Ended	No. of hours	COMPETENCY UNIT Type of Work Performed	Machine/Equipment Details	Name of Trainer/ Supervisor
					Make:	
					Model:	Signature of Supervisor
					Serial No.:	
					Make:	
					Model:	Signature of Supervisor
					Serial No.:	
					Make:	
					Model:	Signature of Supervisor
					Serial No.:	

Date	Time Started	Time Ended	No. of hours	COMPETENCY UNIT Type of Work Performed	Machine/Equipment Details		Name of Trainer/ Supervisor
					Make:		
					Model:		Signature of Supervisor
					Serial No.:		
					Make:		
					Model:		Signature of Supervisor
					Serial No.:		
					Make:		
					Model:		Signature of Supervisor
					Serial No.:		

Date	Time Started	Time Ended	No. of hours	COMPETENCY UNIT Type of Work Performed	Machine/Equipment Details	Name of Trainer/ Supervisor
					Make:	
					Model:	Signature of Supervisor
					Serial No.:	
					Make:	
					Model:	Signature of Supervisor
					Serial No.:	
					Make:	
					Model:	Signature of Supervisor
					Serial No.:	

Date	Time Started	Time Ended	No. of hours	COMPETENCY UNIT Type of Work Performed	Machine/Equipment Details	Name of Trainer/ Supervisor
					Make:	
					Model:	Signature of Supervisor
					Serial No.:	
					Make:	
					Model:	Signature of Supervisor
					Serial No.:	
					Make:	
					Model:	Signature of Supervisor
					Serial No.:	

Date	Time Started	Time Ended	No. of hours	COMPETENCY UNIT Type of Work Performed	Machine/Equipment Details	Name of Trainer/ Supervisor
					Make:	
					Model:	**Signature of Supervisor**
					Serial No.:	
					Make:	
					Model:	**Signature of Supervisor**
					Serial No.:	
					Make:	
					Model:	**Signature of Supervisor**
					Serial No.:	

Date	Time Started	Time Ended	No. of hours	COMPETENCY UNIT Type of Work Performed	Machine/Equipment Details	Name of Trainer/ Supervisor
					Make:	
					Model:	Signature of Supervisor
					Serial No.:	
					Make:	
					Model:	Signature of Supervisor
					Serial No.:	
					Make:	
					Model:	Signature of Supervisor
					Serial No.:	

Date	Time Started	Time Ended	No. of hours	COMPETENCY UNIT Type of Work Performed	Machine/Equipment Details		Name of Trainer/ Supervisor
					Make:		
					Model:		Signature of Supervisor
					Serial No.:		
					Make:		
					Model:		Signature of Supervisor
					Serial No.:		
					Make:		
					Model:		Signature of Supervisor
					Serial No.:		

Date	Time Started	Time Ended	No. of hours	COMPETENCY UNIT Type of Work Performed	Machine/Equipment Details	Name of Trainer/ Supervisor
					Make:	
					Model:	Signature of Supervisor
					Serial No.:	
					Make:	
					Model:	Signature of Supervisor
					Serial No.:	
					Make:	
					Model:	Signature of Supervisor
					Serial No.:	

Date	Time Started	Time Ended	No. of hours	COMPETENCY UNIT Type of Work Performed	Machine/Equipment Details	Name of Trainer/ Supervisor
					Make:	
					Model:	Signature of Supervisor
					Serial No.:	
					Make:	
					Model:	Signature of Supervisor
					Serial No.:	
					Make:	
					Model:	Signature of Supervisor
					Serial No.:	

Date	Time Started	Time Ended	No. of hours	COMPETENCY UNIT Type of Work Performed	Machine/Equipment Details		Name of Trainer/ Supervisor
					Make:		
					Model:		Signature of Supervisor
					Serial No.:		
					Make:		
					Model:		Signature of Supervisor
					Serial No.:		
					Make:		
					Model:		Signature of Supervisor
					Serial No.:		

Date	Time Started	Time Ended	No. of hours	**COMPETENCY UNIT** **Type of Work Performed**	Machine/Equipment Details	Name of Trainer/ Supervisor
					Make:	
					Model:	**Signature of Supervisor**
					Serial No.:	
					Make:	
					Model:	**Signature of Supervisor**
					Serial No.:	
					Make:	
					Model:	**Signature of Supervisor**
					Serial No.:	

Date	Time Started	Time Ended	No. of hours	COMPETENCY UNIT Type of Work Performed	Machine/Equipment Details	Name of Trainer/ Supervisor
					Make:	
					Model:	Signature of Supervisor
					Serial No.:	
					Make:	
					Model:	Signature of Supervisor
					Serial No.:	
					Make:	
					Model:	Signature of Supervisor
					Serial No.:	

Date	Time Started	Time Ended	No. of hours	COMPETENCY UNIT Type of Work Performed	Machine/Equipment Details	Name of Trainer/ Supervisor
					Make:	
					Model:	Signature of Supervisor
					Serial No.:	
					Make:	
					Model:	Signature of Supervisor
					Serial No.:	
					Make:	
					Model:	Signature of Supervisor
					Serial No.:	

Date	Time Started	Time Ended	No. of hours	COMPETENCY UNIT Type of Work Performed	Machine/Equipment Details	Name of Trainer/ Supervisor
					Make:	
					Model:	Signature of Supervisor
					Serial No.:	
					Make:	
					Model:	Signature of Supervisor
					Serial No.:	
					Make:	
					Model:	Signature of Supervisor
					Serial No.:	

Date	Time Started	Time Ended	No. of hours	COMPETENCY UNIT Type of Work Performed	Machine/Equipment Details	Name of Trainer/ Supervisor
					Make:	
					Model:	Signature of Supervisor
					Serial No.:	
					Make:	
					Model:	Signature of Supervisor
					Serial No.:	
					Make:	
					Model:	Signature of Supervisor
					Serial No.:	

Date	Time Started	Time Ended	No. of hours	COMPETENCY UNIT Type of Work Performed	Machine/Equipment Details	Name of Trainer/ Supervisor
					Make:	
					Model:	**Signature of Supervisor**
					Serial No.:	
					Make:	
					Model:	**Signature of Supervisor**
					Serial No.:	
					Make:	
					Model:	**Signature of Supervisor**
					Serial No.:	

Date	Time Started	Time Ended	No. of hours	COMPETENCY UNIT Type of Work Performed	Machine/Equipment Details	Name of Trainer/ Supervisor
					Make:	
					Model:	Signature of Supervisor
					Serial No.:	
					Make:	
					Model:	Signature of Supervisor
					Serial No.:	
					Make:	
					Model:	Signature of Supervisor
					Serial No.:	

Date	Time Started	Time Ended	No. of hours	COMPETENCY UNIT Type of Work Performed	Machine/Equipment Details	Name of Trainer/ Supervisor
					Make:	
					Model:	Signature of Supervisor
					Serial No.:	
					Make:	
					Model:	Signature of Supervisor
					Serial No.:	
					Make:	
					Model:	Signature of Supervisor
					Serial No.:	

Date	Time Started	Time Ended	No. of hours	**COMPETENCY UNIT** **Type of Work Performed**	Machine/Equipment Details	Name of Trainer/ Supervisor
					Make:	
					Model:	**Signature of Supervisor**
					Serial No.:	
					Make:	
					Model:	**Signature of Supervisor**
					Serial No.:	
					Make:	
					Model:	**Signature of Supervisor**
					Serial No.:	

Date	Time Started	Time Ended	No. of hours	COMPETENCY UNIT Type of Work Performed	Machine/Equipment Details	Name of Trainer/ Supervisor
					Make:	
					Model:	Signature of Supervisor
					Serial No.:	
					Make:	
					Model:	Signature of Supervisor
					Serial No.:	
					Make:	
					Model:	Signature of Supervisor
					Serial No.:	

Date	Time Started	Time Ended	No. of hours	**COMPETENCY UNIT** **Type of Work Performed**	**Machine/Equipment Details**	**Name of Trainer/ Supervisor**
					Make:	
					Model:	**Signature of Supervisor**
					Serial No.:	
					Make:	
					Model:	**Signature of Supervisor**
					Serial No.:	
					Make:	
					Model:	**Signature of Supervisor**
					Serial No.:	

Date	Time Started	Time Ended	No. of hours	COMPETENCY UNIT Type of Work Performed	Machine/Equipment Details	Name of Trainer/ Supervisor
					Make:	
					Model:	Signature of Supervisor
					Serial No.:	
					Make:	
					Model:	Signature of Supervisor
					Serial No.:	
					Make:	
					Model:	Signature of Supervisor
					Serial No.:	

Date	Time Started	Time Ended	No. of hours	COMPETENCY UNIT Type of Work Performed	Machine/Equipment Details	Name of Trainer/ Supervisor
					Make:	
					Model:	Signature of Supervisor
					Serial No.:	
					Make:	
					Model:	Signature of Supervisor
					Serial No.:	
					Make:	
					Model:	Signature of Supervisor
					Serial No.:	

Date	Time Started	Time Ended	No. of hours	COMPETENCY UNIT Type of Work Performed	Machine/Equipment Details	Name of Trainer/ Supervisor
					Make:	
					Model:	Signature of Supervisor
					Serial No.:	
					Make:	
					Model:	Signature of Supervisor
					Serial No.:	
					Make:	
					Model:	Signature of Supervisor
					Serial No.:	

Date	Time Started	Time Ended	No. of hours	COMPETENCY UNIT Type of Work Performed	Machine/Equipment Details	Name of Trainer/ Supervisor
					Make:	
					Model:	Signature of Supervisor
					Serial No.:	
					Make:	
					Model:	Signature of Supervisor
					Serial No.:	
					Make:	
					Model:	Signature of Supervisor
					Serial No.:	

Date	Time Started	Time Ended	No. of hours	COMPETENCY UNIT Type of Work Performed	Machine/Equipment Details	Name of Trainer/ Supervisor
					Make:	
					Model:	Signature of Supervisor
					Serial No.:	
					Make:	
					Model:	Signature of Supervisor
					Serial No.:	
					Make:	
					Model:	Signature of Supervisor
					Serial No.:	

Date	Time Started	Time Ended	No. of hours	COMPETENCY UNIT Type of Work Performed	Machine/Equipment Details	Name of Trainer/ Supervisor
					Make:	
					Model:	Signature of Supervisor
					Serial No.:	
					Make:	
					Model:	Signature of Supervisor
					Serial No.:	
					Make:	
					Model:	Signature of Supervisor
					Serial No.:	

Date	Time Started	Time Ended	No. of hours	COMPETENCY UNIT Type of Work Performed	Machine/Equipment Details	Name of Trainer/ Supervisor
					Make:	
					Model:	**Signature of Supervisor**
					Serial No.:	
					Make:	
					Model:	**Signature of Supervisor**
					Serial No.:	
					Make:	
					Model:	**Signature of Supervisor**
					Serial No.:	

Date	Time Started	Time Ended	No. of hours	COMPETENCY UNIT Type of Work Performed	Machine/Equipment Details	Name of Trainer/ Supervisor
					Make:	
					Model:	Signature of Supervisor
					Serial No.:	
					Make:	
					Model:	Signature of Supervisor
					Serial No.:	
					Make:	
					Model:	Signature of Supervisor
					Serial No.:	

Date	Time Started	Time Ended	No. of hours	COMPETENCY UNIT Type of Work Performed	Machine/Equipment Details	Name of Trainer/ Supervisor
					Make:	
					Model:	Signature of Supervisor
					Serial No.:	
					Make:	
					Model:	Signature of Supervisor
					Serial No.:	
					Make:	
					Model:	Signature of Supervisor
					Serial No.:	

Date	Time Started	Time Ended	No. of hours	COMPETENCY UNIT Type of Work Performed	Machine/Equipment Details	Name of Trainer/ Supervisor
					Make:	
					Model:	Signature of Supervisor
					Serial No.:	
					Make:	
					Model:	Signature of Supervisor
					Serial No.:	
					Make:	
					Model:	Signature of Supervisor
					Serial No.:	

Date	Time Started	Time Ended	No. of hours	COMPETENCY UNIT Type of Work Performed	Machine/Equipment Details	Name of Trainer/ Supervisor
					Make:	
					Model:	Signature of Supervisor
					Serial No.:	
					Make:	
					Model:	Signature of Supervisor
					Serial No.:	
					Make:	
					Model:	Signature of Supervisor
					Serial No.:	

Date	Time Started	Time Ended	No. of hours	COMPETENCY UNIT Type of Work Performed	Machine/Equipment Details	Name of Trainer/ Supervisor
					Make:	
					Model:	Signature of Supervisor
					Serial No.:	
					Make:	
					Model:	Signature of Supervisor
					Serial No.:	
					Make:	
					Model:	Signature of Supervisor
					Serial No.:	

Date	Time Started	Time Ended	No. of hours	COMPETENCY UNIT Type of Work Performed	Machine/Equipment Details		Name of Trainer/ Supervisor
					Make:		
					Model:		Signature of Supervisor
					Serial No.:		
					Make:		
					Model:		Signature of Supervisor
					Serial No.:		
					Make:		
					Model:		Signature of Supervisor
					Serial No.:		

Date	Time Started	Time Ended	No. of hours	COMPETENCY UNIT Type of Work Performed	Machine/Equipment Details	Name of Trainer/ Supervisor
					Make:	
					Model:	**Signature of Supervisor**
					Serial No.:	
					Make:	
					Model:	**Signature of Supervisor**
					Serial No.:	
					Make:	
					Model:	**Signature of Supervisor**
					Serial No.:	

Date	Time Started	Time Ended	No. of hours	COMPETENCY UNIT Type of Work Performed	Machine/Equipment Details	Name of Trainer/ Supervisor
					Make:	
					Model:	Signature of Supervisor
					Serial No.:	
					Make:	
					Model:	Signature of Supervisor
					Serial No.:	
					Make:	
					Model:	Signature of Supervisor
					Serial No.:	

Date	Time Started	Time Ended	No. of hours	COMPETENCY UNIT Type of Work Performed	Machine/Equipment Details	Name of Trainer/ Supervisor
					Make:	
					Model:	Signature of Supervisor
					Serial No.:	
					Make:	
					Model:	Signature of Supervisor
					Serial No.:	
					Make:	
					Model:	Signature of Supervisor
					Serial No.:	

Date	Time Started	Time Ended	No. of hours	COMPETENCY UNIT Type of Work Performed	Machine/Equipment Details	Name of Trainer/ Supervisor
					Make:	
					Model:	Signature of Supervisor
					Serial No.:	
					Make:	
					Model:	Signature of Supervisor
					Serial No.:	
					Make:	
					Model:	Signature of Supervisor
					Serial No.:	

Date	Time Started	Time Ended	No. of hours	COMPETENCY UNIT Type of Work Performed	Machine/Equipment Details	Name of Trainer/ Supervisor
					Make:	
					Model:	**Signature of Supervisor**
					Serial No.:	
					Make:	
					Model:	**Signature of Supervisor**
					Serial No.:	
					Make:	
					Model:	**Signature of Supervisor**
					Serial No.:	

Date	Time Started	Time Ended	No. of hours	COMPETENCY UNIT Type of Work Performed	Machine/Equipment Details	Name of Trainer/ Supervisor
					Make:	
					Model:	Signature of Supervisor
					Serial No.:	
					Make:	
					Model:	Signature of Supervisor
					Serial No.:	
					Make:	
					Model:	Signature of Supervisor
					Serial No.:	

Date	Time Started	Time Ended	No. of hours	COMPETENCY UNIT Type of Work Performed	Machine/Equipment Details	Name of Trainer/ Supervisor
					Make:	
					Model:	Signature of Supervisor
					Serial No.:	
					Make:	
					Model:	Signature of Supervisor
					Serial No.:	
					Make:	
					Model:	Signature of Supervisor
					Serial No.:	

Date	Time Started	Time Ended	No. of hours	COMPETENCY UNIT Type of Work Performed	Machine/Equipment Details	Name of Trainer/ Supervisor
					Make:	
					Model:	Signature of Supervisor
					Serial No.:	
					Make:	
					Model:	Signature of Supervisor
					Serial No.:	
					Make:	
					Model:	Signature of Supervisor
					Serial No.:	

Date	Time Started	Time Ended	No. of hours	COMPETENCY UNIT Type of Work Performed	Machine/Equipment Details	Name of Trainer/ Supervisor
					Make:	
					Model:	**Signature of Supervisor**
					Serial No.:	
					Make:	
					Model:	**Signature of Supervisor**
					Serial No.:	
					Make:	
					Model:	**Signature of Supervisor**
					Serial No.:	

Date	Time Started	Time Ended	No. of hours	COMPETENCY UNIT Type of Work Performed	Machine/Equipment Details	Name of Trainer/ Supervisor
					Make:	
					Model:	Signature of Supervisor
					Serial No.:	
					Make:	
					Model:	Signature of Supervisor
					Serial No.:	
					Make:	
					Model:	Signature of Supervisor
					Serial No.:	

Date	Time Started	Time Ended	No. of hours	COMPETENCY UNIT Type of Work Performed	Machine/Equipment Details	Name of Trainer/ Supervisor
					Make:	
					Model:	Signature of Supervisor
					Serial No.:	
					Make:	
					Model:	Signature of Supervisor
					Serial No.:	
					Make:	
					Model:	Signature of Supervisor
					Serial No.:	

Date	Time Started	Time Ended	No. of hours	COMPETENCY UNIT Type of Work Performed	Machine/Equipment Details	Name of Trainer/ Supervisor
					Make:	
					Model:	Signature of Supervisor
					Serial No.:	
					Make:	
					Model:	Signature of Supervisor
					Serial No.:	
					Make:	
					Model:	Signature of Supervisor
					Serial No.:	

Date	Time Started	Time Ended	No. of hours	COMPETENCY UNIT Type of Work Performed	Machine/Equipment Details	Name of Trainer/ Supervisor
					Make:	
					Model:	Signature of Supervisor
					Serial No.:	
					Make:	
					Model:	Signature of Supervisor
					Serial No.:	
					Make:	
					Model:	Signature of Supervisor
					Serial No.:	

Date	Time Started	Time Ended	No. of hours	COMPETENCY UNIT Type of Work Performed	Machine/Equipment Details	Name of Trainer/ Supervisor
					Make:	
					Model:	Signature of Supervisor
					Serial No.:	
					Make:	
					Model:	Signature of Supervisor
					Serial No.:	
					Make:	
					Model:	Signature of Supervisor
					Serial No.:	

Date	Time Started	Time Ended	No. of hours	COMPETENCY UNIT Type of Work Performed	Machine/Equipment Details	Name of Trainer/ Supervisor
					Make:	
					Model:	Signature of Supervisor
					Serial No.:	
					Make:	
					Model:	Signature of Supervisor
					Serial No.:	
					Make:	
					Model:	Signature of Supervisor
					Serial No.:	

Date	Time Started	Time Ended	No. of hours	COMPETENCY UNIT Type of Work Performed	Machine/Equipment Details	Name of Trainer/ Supervisor
					Make:	
					Model:	Signature of Supervisor
					Serial No.:	
					Make:	
					Model:	Signature of Supervisor
					Serial No.:	
					Make:	
					Model:	Signature of Supervisor
					Serial No.:	

Date	Time Started	Time Ended	No. of hours	COMPETENCY UNIT Type of Work Performed	Machine/Equipment Details	Name of Trainer/ Supervisor
					Make:	
					Model:	Signature of Supervisor
					Serial No.:	
					Make:	
					Model:	Signature of Supervisor
					Serial No.:	
					Make:	
					Model:	Signature of Supervisor
					Serial No.:	

Date	Time Started	Time Ended	No. of hours	COMPETENCY UNIT Type of Work Performed	Machine/Equipment Details	Name of Trainer/ Supervisor
					Make:	
					Model:	Signature of Supervisor
					Serial No.:	
					Make:	
					Model:	Signature of Supervisor
					Serial No.:	
					Make:	
					Model:	Signature of Supervisor
					Serial No.:	

Date	Time Started	Time Ended	No. of hours	COMPETENCY UNIT Type of Work Performed	Machine/Equipment Details	Name of Trainer/ Supervisor
					Make:	
					Model:	**Signature of Supervisor**
					Serial No.:	
					Make:	
					Model:	**Signature of Supervisor**
					Serial No.:	
					Make:	
					Model:	**Signature of Supervisor**
					Serial No.:	

Date	Time Started	Time Ended	No. of hours	COMPETENCY UNIT Type of Work Performed	Machine/Equipment Details	Name of Trainer/ Supervisor
					Make:	
					Model:	Signature of Supervisor
					Serial No.:	
					Make:	
					Model:	Signature of Supervisor
					Serial No.:	
					Make:	
					Model:	Signature of Supervisor
					Serial No.:	

Date	Time Started	Time Ended	No. of hours	COMPETENCY UNIT Type of Work Performed	Machine/Equipment Details	Name of Trainer/ Supervisor
					Make:	
					Model:	Signature of Supervisor
					Serial No.:	
					Make:	
					Model:	Signature of Supervisor
					Serial No.:	
					Make:	
					Model:	Signature of Supervisor
					Serial No.:	

Date	Time Started	Time Ended	No. of hours	COMPETENCY UNIT Type of Work Performed	Machine/Equipment Details	Name of Trainer/ Supervisor
					Make:	
					Model:	Signature of Supervisor
					Serial No.:	
					Make:	
					Model:	Signature of Supervisor
					Serial No.:	
					Make:	
					Model:	Signature of Supervisor
					Serial No.:	

Date	Time Started	Time Ended	No. of hours	COMPETENCY UNIT Type of Work Performed	Machine/Equipment Details	Name of Trainer/ Supervisor
					Make:	
					Model:	Signature of Supervisor
					Serial No.:	
					Make:	
					Model:	Signature of Supervisor
					Serial No.:	
					Make:	
					Model:	Signature of Supervisor
					Serial No.:	

Date	Time Started	Time Ended	No. of hours	COMPETENCY UNIT Type of Work Performed	Machine/Equipment Details	Name of Trainer/ Supervisor
					Make:	
					Model:	Signature of Supervisor
					Serial No.:	
					Make:	
					Model:	Signature of Supervisor
					Serial No.:	
					Make:	
					Model:	Signature of Supervisor
					Serial No.:	

Date	Time Started	Time Ended	No. of hours	COMPETENCY UNIT Type of Work Performed	Machine/Equipment Details	Name of Trainer/ Supervisor
					Make:	
					Model:	Signature of Supervisor
					Serial No.:	
					Make:	
					Model:	Signature of Supervisor
					Serial No.:	
					Make:	
					Model:	Signature of Supervisor
					Serial No.:	

Date	Time Started	Time Ended	No. of hours	COMPETENCY UNIT Type of Work Performed	Machine/Equipment Details		Name of Trainer/ Supervisor
					Make:		
					Model:		Signature of Supervisor
					Serial No.:		
					Make:		
					Model:		Signature of Supervisor
					Serial No.:		
					Make:		
					Model:		Signature of Supervisor
					Serial No.:		

Date	Time Started	Time Ended	No. of hours	COMPETENCY UNIT Type of Work Performed	Machine/Equipment Details		Name of Trainer/ Supervisor
					Make:		
					Model:		Signature of Supervisor
					Serial No.:		
					Make:		
					Model:		Signature of Supervisor
					Serial No.:		
					Make:		
					Model:		Signature of Supervisor
					Serial No.:		

Date	Time Started	Time Ended	No. of hours	COMPETENCY UNIT Type of Work Performed	Machine/Equipment Details	Name of Trainer/ Supervisor
					Make:	
					Model:	Signature of Supervisor
					Serial No.:	
					Make:	
					Model:	Signature of Supervisor
					Serial No.:	
					Make:	
					Model:	Signature of Supervisor
					Serial No.:	

Date	Time Started	Time Ended	No. of hours	COMPETENCY UNIT Type of Work Performed	Machine/Equipment Details		Name of Trainer/ Supervisor
					Make:		
					Model:		Signature of Supervisor
					Serial No.:		
					Make:		
					Model:		Signature of Supervisor
					Serial No.:		
					Make:		
					Model:		Signature of Supervisor
					Serial No.:		

Date	Time Started	Time Ended	No. of hours	COMPETENCY UNIT Type of Work Performed	Machine/Equipment Details	Name of Trainer/ Supervisor
					Make:	
					Model:	Signature of Supervisor
					Serial No.:	
					Make:	
					Model:	Signature of Supervisor
					Serial No.:	
					Make:	
					Model:	Signature of Supervisor
					Serial No.:	

Date	Time Started	Time Ended	No. of hours	COMPETENCY UNIT Type of Work Performed	Machine/Equipment Details	Name of Trainer/ Supervisor
					Make:	
					Model:	Signature of Supervisor
					Serial No.:	
					Make:	
					Model:	Signature of Supervisor
					Serial No.:	
					Make:	
					Model:	Signature of Supervisor
					Serial No.:	

Date	Time Started	Time Ended	No. of hours	COMPETENCY UNIT Type of Work Performed	Machine/Equipment Details	Name of Trainer/ Supervisor
					Make:	
					Model:	Signature of Supervisor
					Serial No.:	
					Make:	
					Model:	Signature of Supervisor
					Serial No.:	
					Make:	
					Model:	Signature of Supervisor
					Serial No.:	

Date	Time Started	Time Ended	No. of hours	COMPETENCY UNIT Type of Work Performed	Machine/Equipment Details	Name of Trainer/ Supervisor
					Make:	
					Model:	**Signature of Supervisor**
					Serial No.:	
					Make:	
					Model:	**Signature of Supervisor**
					Serial No.:	
					Make:	
					Model:	**Signature of Supervisor**
					Serial No.:	

Date	Time Started	Time Ended	No. of hours	COMPETENCY UNIT Type of Work Performed	Machine/Equipment Details		Name of Trainer/ Supervisor
					Make:		
					Model:		Signature of Supervisor
					Serial No.:		
					Make:		
					Model:		Signature of Supervisor
					Serial No.:		
					Make:		
					Model:		Signature of Supervisor
					Serial No.:		

Date	Time Started	Time Ended	No. of hours	COMPETENCY UNIT Type of Work Performed	Machine/Equipment Details	Name of Trainer/ Supervisor
					Make:	
					Model:	Signature of Supervisor
					Serial No.:	
					Make:	
					Model:	Signature of Supervisor
					Serial No.:	
					Make:	
					Model:	Signature of Supervisor
					Serial No.:	

Date	Time Started	Time Ended	No. of hours	COMPETENCY UNIT Type of Work Performed	Machine/Equipment Details	Name of Trainer/ Supervisor
					Make:	
					Model:	Signature of Supervisor
					Serial No.:	
					Make:	
					Model:	Signature of Supervisor
					Serial No.:	
					Make:	
					Model:	Signature of Supervisor
					Serial No.:	

Date	Time Started	Time Ended	No. of hours	COMPETENCY UNIT Type of Work Performed	Machine/Equipment Details	Name of Trainer/ Supervisor
					Make:	
					Model:	Signature of Supervisor
					Serial No.:	
					Make:	
					Model:	Signature of Supervisor
					Serial No.:	
					Make:	
					Model:	Signature of Supervisor
					Serial No.:	

Date	Time Started	Time Ended	No. of hours	COMPETENCY UNIT Type of Work Performed	Machine/Equipment Details	Name of Trainer/ Supervisor
					Make:	
					Model:	Signature of Supervisor
					Serial No.:	
					Make:	
					Model:	Signature of Supervisor
					Serial No.:	
					Make:	
					Model:	Signature of Supervisor
					Serial No.:	

Date	Time Started	Time Ended	No. of hours	COMPETENCY UNIT Type of Work Performed	Machine/Equipment Details	Name of Trainer/ Supervisor
					Make:	
					Model:	**Signature of Supervisor**
					Serial No.:	
					Make:	
					Model:	**Signature of Supervisor**
					Serial No.:	
					Make:	
					Model:	**Signature of Supervisor**
					Serial No.:	

Date	Time Started	Time Ended	No. of hours	COMPETENCY UNIT Type of Work Performed	Machine/Equipment Details	Name of Trainer/ Supervisor
					Make:	
					Model:	Signature of Supervisor
					Serial No.:	
					Make:	
					Model:	Signature of Supervisor
					Serial No.:	
					Make:	
					Model:	Signature of Supervisor
					Serial No.:	

Date	Time Started	Time Ended	No. of hours	COMPETENCY UNIT Type of Work Performed	Machine/Equipment Details	Name of Trainer/ Supervisor
					Make:	
					Model:	**Signature of Supervisor**
					Serial No.:	
					Make:	
					Model:	**Signature of Supervisor**
					Serial No.:	
					Make:	
					Model:	**Signature of Supervisor**
					Serial No.:	

Date	Time Started	Time Ended	No. of hours	COMPETENCY UNIT Type of Work Performed	Machine/Equipment Details	Name of Trainer/ Supervisor
					Make:	
					Model:	Signature of Supervisor
					Serial No.:	
					Make:	
					Model:	Signature of Supervisor
					Serial No.:	
					Make:	
					Model:	Signature of Supervisor
					Serial No.:	

Date	Time Started	Time Ended	No. of hours	COMPETENCY UNIT Type of Work Performed	Machine/Equipment Details	Name of Trainer/ Supervisor
					Make:	
					Model:	Signature of Supervisor
					Serial No.:	
					Make:	
					Model:	Signature of Supervisor
					Serial No.:	
					Make:	
					Model:	Signature of Supervisor
					Serial No.:	

Date	Time Started	Time Ended	No. of hours	COMPETENCY UNIT Type of Work Performed	Machine/Equipment Details		Name of Trainer/ Supervisor
					Make:		
					Model:		**Signature of Supervisor**
					Serial No.:		
					Make:		
					Model:		**Signature of Supervisor**
					Serial No.:		
					Make:		
					Model:		**Signature of Supervisor**
					Serial No.:		

Made in the USA
Monee, IL
06 December 2022